Emerging Technologies in Clean Energy

Topical Conference at the 2021 AIChE Spring Meeting and 17th Global Congress on Process Safety

Online
18 – 23 April 2021

ISBN: 978-1-7138-5330-5

Emerging Technologies in Clean Energy

Topical Conference at the 2021 AIChE Spring Meeting and 17th Global Congress on Process Safety

Online
18 – 23 April 2021

AIChE
The Global Home of Chemical Engineers

Printed from e-media with permission by:

Curran Associates, Inc.
57 Morehouse Lane
Red Hook, NY 12571

AIChE
120 Wall Street, FL 23
New York, NY 10005-4020

Phone: (800) 242-4363
Fax: (203) 775-5177

www.aiche.org

Additional copies of this publication are available from:

Curran Associates, Inc.
57 Morehouse Lane
Red Hook, NY 12571 USA
Phone: 845-758-0400
Fax: 845-758-2633
Email: curran@proceedings.com
Web: www.proceedings.com

TABLE OF CONTENTS

618983 Thermal Energy Production and Heat Exchange between an Electrochemical Cell and Its Surroundings

Abstract:

Information regarding the heat exchange between an electrochemical cell and its surroundings is of great utility for the planet earth and deep space probe applications. Thermal energy production must be controlled in an electrochemical cell to avoid thermal runaway and cell rupture. For the operation of a cell at a desired temperature, even under reversible operational conditions, the generated thermal energy must be removed as heat transfer from the cell interior to its surroundings. Here, a theoretical scheme has been developed to determine the cell temperature for any desired cell discharge current. This scheme incorporates both the reversible production of thermal energy due to the change in entropy of the system and the irreversible production of thermal energy due to the cell voltage losses associated with the species transport in the cell electrolyte, electrode components, current collectors, and the electrochemical reactions involving charge transfer at the electrolyte-electrode interfaces. Assumptions for this model include that natural convection heat transfer may be neglected in a perfect vacuum and that the cell temperature throughout its body is constant at any time during the cell discharge. The information on dT/dt versus t (i.e., rate of change of the cell temperature versus time) generated by this scheme can be integrated either graphically or by the use of a numerical procedure, such as the Euler method, to obtain the cell temperature versus time data. Then, plots of temperature versus time can be prepared for each desired current value. Currently, we are in the process of calculating temperature versus time data. It is anticipated that such data will become available in the near future.

619084 Developing Sustainable, Safer and Integrated Green H_2 Supply Chains for Qatar's Industrial,Transportation and Shipping Sector

Abstract:

Keywords: Decarbonization, Green Hydrogen; Supply chain optimization, Techno-economic-environmental analysis;

Hydrogen (H_2) can play a significant role in the global energy mix (IEA, 2019). Hydrogen based economy can be a great opportunity for a country like Qatar to decarbonize its multiple sectors including transportation, shipping, global energy markets and industrial sectors. Qatar has the industrial portfolio, has access to the resources, the commitment and the vision to become the center of the hydrogen supply chain network (HSCN). It will provide a great opportunity for the country to decarbonize its energy and industry sector, greatly reduce its Greenhouse gas emission, and improving its air quality. The use of green or blue H_2 for the production of value-added products in Qatar will contribute greatly to growing Qatar's industrial portfolio, an added economic value to the State of Qatar. According to the Hydrogen Council, a United Nations entity under the World Economic Forum, Hydrogen is projected to contribute greatly to the transportation sector (around 150 MTPA), industrial sector (around 110 MTPA), followed by power generation (around 140 MTPA) and others (Hydrogen Council, 2017). Hence, large scale production of $H_2$from cleaner sources is essential to meet the rising demand for H_2 in transportation and all sectors.

However, there are still some barriers to the realization of a hydrogen-based economy, which includes large scale hydrogen production cost, infrastructure investments, bulk storage, transport & distribution, safety consideration, and matching supply-demand uncertainties (Kazi et al., 2020). This work highlights how the aforementioned challenges can be handled strategically through a multi-sector industrial-urban symbiosis for the hydrogen supply chain implementation. Such symbiosis can enhance the mutual relationship between diverse industries and urban planning by exploring varied scopes of multi-purpose hydrogen usage (i.e., clean energy source as a safer carrier, industrial feedstock and intermittent products, vehicle and shipping fuel, international energy trading, etc.) both in local and international markets. It enables individual entities and businesses to participate in the physical exchange of materials, by-products, energy, and water, with strategic advantages for all participants. Besides, waste/by-product exchanges, several different kinds of synergies are also possible, such as the sharing of resources and shared facilities.

This study focuses on the development of a strategic framework for the design of a hydrogen supply chain network (HSCN) mainly investigating the potential of industrial decarbonization and multi-sectors integration (i.e., transportation, energy, shipping) via green hydrogen economy. The problem was formulated as a mixed integer linear programming (MILP) and solved in GAMS/ IBM ILOG CPLEX 30.3.0 solver. The applicability of the developed model was demonstrated using a base case Eco-industrial city consisting of 10 diverse industrial portfolios targeting decarbonization by 5%. The solution was able to find the optimal HSCN from the probable superstructure along with the optimal sizing of green hydrogen production (453.03 MM kg/y), optimal water sources, optimal sinks and the optimal amount of byproducts generation. Furthermore, the multi-purpose model can accomplish detailed techno-economic-environmental analysis for variable scenarios (e.g., variation in decarbonization target, liquid hydrogen demand, hydrogen production cost, earnings from byproducts) based on net present value. The concept and the formulation presented in this manuscript will act as the building block for future complex supply chain problems such as crossover of the multi-sector supply chains, the impact of time-varying and multiple storages to manage real-time supply chain disruption, varying cost of unit hydrogen production to accommodate the rapid advancement of technology, etc.

References

Hydrogen Council. (2017). Hydrogen scaling up. https://hydrogencouncil.com/wp-content/uploads/2017/11/Hydrogen-scaling-up-Hydrogen-Council.pdf.

IEA. (2019). World Energy Outlook 2019. Flagship report, https://www.iea.org/reports/world-energy-outlook-2019.

Kazi, M.-K., Eljack, F., El-Halwagi, M. M., & Haouari, M. (2020). Green hydrogen for industrial sector decarbonization: Costs and impacts on hydrogen economy in qatar. Computers & Chemical Engineering, 107144.

618935 Revolutionary CANS Catalyst Carrier Technology for Fischer-Tropsch Synthesis: A Sustainable Approach to Fuels Production

Abstract:

A sustained effort is required over the next few decades to halt climate change due to global warming by reducing CO_2 emissions. The increasing availability of renewable electricity and development of new technology such as high energy density batteries and fuel cells will reduce our reliance on fossil fuels; however, some applications, such as long-haul aviation, are particularly challenging to de-carbonise.

Sustainable liquid fuels which fit into exiting supply chains and avoid changes to transportation infrastructure and engines are easiest to adopt. The conversion of waste, biomass or existing CO_2 emissions into sustainable fuels via Fischer-Tropsch synthesis offers one solution to this problem.

JM/BP have been working together for many years on Fischer-Tropsch technology culminating in the development of the revolutionary **CANS**^{TM} catalyst carrier technology, which enables sustainable aviation and other liquid fuels to be made economically at a wide range of production capacities.

This technology was selected by Fulcrum Bioenergy for its first commercial-scale waste to liquid fuels facility in Storey County, Nevada, USA.

We will describe the benefits of the **CANS** catalyst carrier technology and how the various aspects of the technology have been scaled up from the laboratory to full scale manufacture.

619192 Two Stage Anaerobic Digestion with Internal Heat Recycle

Abstract:

Anaerobic digestion of confined agricultural animals waste can lead the way to reduction in equivalent carbon dioxide emissions to pre 1980 levels. A general rule of thumb for ambient anaerobic digesters in northern areas of the US is that they do not work well north of interstate 70 in the summer and not at all in the winter. New two stage anaerobic digestion with internal heat recycle makes the systems function and profitable in the northern areas of the US as well as the south.

This paper will explain how new patented technology functions increases gas output, improves gas quality and rate.

618840 Formation of Side Products in the Synthesis of the Synthetic Fuel Poly(oxymethylene) Dimethyl Ethers and the Implications for Process Design

Abstract:

A substantial reduction of greenhouse gas emissions in each economic sector is essential in order to limit global warming. Especially the transportation sector lacks of quickly deployable alternative technologies due to the wide-spread use of internal combustion engines. Poly(oxymethylene) dimethyl ethers (OME) can be used as fuel in conventional diesel engines, offer excellent fuel properties like soot-free combustion and also reduce local NOx emissions [1]. Additionally, they can contribute to a sustainable transportation sector since they are based on renewable carbon and energy sources [2].

Several synthesis routes for the production of OME usually starting from methanol or dimethyl ether as feedstock are known. Methanol is converted to formaldehyde, which is then reacted with further methanol or dimethyl ether to OME in an acid-catalyzed equilibrium reaction. A recently developed water-tolerant OME process concept uses a methanolic formaldehyde solution as educt and, thus, avoids expensive intermediate process steps [3]. The main reactions leading to the formation of OME are well-studied in literature [4].

This presentation discusses the routes toward OME comprehensively. It is highlighted which lines of research are currently done in Europe. Further, novel experimental results for the formation of the most relevant side products (trioxane, methyl formate and formic acid) during the synthesis of OME from methanolic formaldehyde solutions under industrial conditions are presented. The acidic ion exchange resin Amberlyst 46 was thereby used as heterogeneous catalyst. Long-term experiments under elevated temperature and pressure were carried out in a batch reactor and the influence of the reaction parameters temperature, educt ratio, water concentration and catalyst amount was studied. A developed pseudo-homogeneous kinetic model shows good agreement with the experimental profiles.

Based on the model, the implications for process design are evaluated within a process simulation. All three side products, trioxane, methyl formate, and formic acid, accumulate within the process. Unlike the concentration of trioxane that is limited to chemical equilibrium, the concentrations of methyl formate and formic acid increase steadily and, therefore, a concept for their removal is needed. Considering a simple purge without further purification, a trade-off between reactor temperature, or catalyst amount, and the resulting loss of product in the purge is detected. An optimal operating point for the process is identified.

References

[1] M. Härtl, P. Seidenspinner, E. Jacob, G. Wachtmeister, *Fuel* **2015**, 153, 328-335. DOI: 10.1016/j.fuel.2015.03.012

[2] S. Deutz, D. Bongartz, B. Heuser, A. Kätelhön, L. Schulze Langenhorst, A. Omari, M. Walters, J. Klankermayer, W. Leitner, A. Mitsos, S. Pischinger, A. Bardow, *Energy Environ. Sci.* **2018**, *11* (2), 331-343. DOI: 10.1039/C7EE01657C

[3] N. Schmitz, E. Ströfer, J. Burger, H. Hasse, *Ind. Eng. Chem. Res.* **2017**, 56, 11519–11530. DOI:

10.1021/acs.iecr.7b02314

[4] a) N. Schmitz, J. Burger, H. Hasse, *Ind. Eng. Chem. Res.* **2015** *54* (50), 12553-1256. DOI: 10.1021/acs.iecr.5b04046. b) D. Oestreich, L. Lautenschütz, U. Arnold, J. Sauer, *Chem. Eng. Sci.* **2017,** 163, 92-104. DOI: 10.1016/j.ces.2016.12.037

619160 Novel Fluropolymer Membrane for the Separation of Carbon Dioxide from Methane in Biogas Digestors

Abstract:

Separating carbon dioxide from methane is a key challenge for the renewable gas industry. Biogas is produced largely on farms or small waste centers, not at large centralized plants. While there are existing technologies to separate methane from carbon dioxide, they require multiple stage high pressure, and are sensitive to process poisons. An efficient and cost effective solution is needed to bring down barriers that previously prevented small scale operations from transforming their biogas into usable fuel. CMS has developed a membrane aimed at allowing users to capture, separate and upgrade existing biogas streams that would otherwise be lost or under-valued.

Designed to operate at low pressure while maintaining high selectivity(>20) and permeance, Optiperm™ *biogas* has demonstrated promising performance in laboratory testing. Optiperm™ *biogas* is stable in the presence of water and hydrogen sulfide, requiring less pretreatment. Initial laboratory tests show that the membrane is stable in feeds with up to 100% humidity and 300ppm of hydrogen sulfide.

Results from the first field demonstration of Optiperm™ *biogas* will be shared. Working with Energy Projects Manager Matt Steiman at the Dickinson College Farm, a small spiral wound module system has been installed. The system takes raw biogas feed directly from the digester with no pretreatment. The pilot is intended to demonstrate successful operation under field conditions for an extended period of time. Optiperm™ *biogas* technology will be tested for a period of at least 2 months to demonstrate stability in the field and validate performance measured in the CMS laboratory. Optiperm™ *biogas* upgrading streams from 60% to 90% methane in a single stage.

619220 Black Swan "Wig" Membrane Has Lowest Cost Oxygen Production from Air By Using No Pressurization

Abstract:

The patent pending Black Swan, LLC “Wig" Membranes and Membrane Air Enrichment (MAE) Process have several distinct advantages over other membranes and methods of producing enriched (concentrations above 20.9%) oxygen from air, including highly enriched (90%+) oxygen. The membrane has a novel configuration that utilizes conventional hollow fiber membranes having a 40+ year history. The configuration of the membrane reduces manufacturing cost by almost 1/3, reduces power consumption by 1/3, and increases performance of the membrane by 20%. The MAE process allows for applications for oxygen enrichment at low levels if needed in moderate oxygen enrichment services (under 50%) in a single stage system, as well as more severe services with up to 95% + oxygen enrichment in three stage systems. For systems looking to produce capturable high concentrations (90%+) of CO_2 in stack gas, say of boilers or power plants, the system can be “plug and play” or “drop in”, meaning no requirements to modify the burner or refractory of the fired system. Data from prototypes of two large hollow fiber membrane manufacturers will be presented, as well as pilot plant applications in refinery sulfur recovery units, sanitation aerations systems, and biogas power plants.

619101 Coal Chemical Looping Combustion Using Bimetallic Cu-Fe Oxygen Carrier in a Carbon Dioxide Gas Environment

Abstract:

Chemical-looping combustion (CLC) is mainly proposed for a circulating fluidized bed (CFB) using various fluidizing gases such as carbon dioxide (CO_2) or steam (H_2O). CO_2 in the CLC is also a reactant and the process may be considered as CO_2 utilization. Prior studies showed that bimetallic Cu-Fe oxygen carriers (OC) had higher reactivities compared to monometallic iron oxide. In this study, the Cu-Fe OCs with an inert support silicon dioxide (SiO_2) or zirconium dioxide (ZrO_2) were synthesized using high pressure pelletization followed by calcination at 950 or 1100°C. Reactivities of the OCs with coal char and their CO_2 conversion efficiency were investigated at 950 and 1050°C in multiple-cycle tests under a CO_2 gas environment. The OCs with a ZrO_2 support can be tested at a higher temperature than those with an SiO_2 support due to problems associated with sintering of SiO_2-supported OCs when calcined at 1100°C. Results show high carbon conversions, carbon conversion rates, CO_2 conversion efficiencies and cyclabilities of the combined Cu-Fe OCs with the char under CO_2.

SPRING21
+17TH GCPS
A Joint AIChE and CCPS Meeting

Determining the Effectiveness of the Human Safeguard

James Topp
Provenance Consulting, a Trinity Consultants Company
15995 N. Barkers Landing Rd.
Suite 145
Houston, TX 77079
james.topp@provpsm.com

Prepared for Presentation at
American Institute of Chemical Engineers
2020 Spring Meeting and 17th Global Congress on Process Safety
Virtual
April 18 - 22, 2021

Determining the Effectiveness of the Human Safeguard

James Topp
Provenance Consulting, a Trinity Consultants Company
15995 N. Barkers Landing Rd.
Suite 145
Houston, TX 77079
james.topp@provpsm.com

Keywords: alarms, operator, PHA, human intervention, safeguards, HAZOP, process safety time, human

Abstract

A team conducting a Process Hazard Analysis (PHA) reviews a vessel with a pressure alarm– a common safeguard against many overpressure scenarios the vessel could experience. Everyone on the PHA team feels the alarm provides plenty of time for an operator to respond to any given overpressure scenario and continues with the PHA. This may be acceptable in some cases: the expected response time to that pressure alarm is sufficient for an operator to respond to a blocked outlet. But it could be a costly mistake in other cases: if a bypass on the fuel gas line is left open, the MAWP of the vessel is exceeded in under 60 seconds – not nearly enough time for the operator to respond before potential loss of containment.

This situation – where alarms are used as safeguards without proper vetting and consideration of operator response time to those alarms – is occurring more frequently throughout industry. With the move toward more technology-driven and interconnected facilities, alarms have become one of the most prevalent safeguards in facilities. To appropriately use alarms as safeguards, sites must ensure alarms have clear instructions on the correct response and there is adequate time to mitigate the scenario.

Without a clear idea of which alarms can or cannot be acceptable safeguards for given conditions or scenarios, facilities may leave themselves unprotected for scenarios deemed appropriately mitigated by a risk study. Alternatively, they could unnecessarily spend their budget installing more robust safety systems when operator intervention through alarm would be adequate.

This paper will discuss ways to determine the amount of time an operator would have to respond to an alarm condition, ways to account for the human factors of alarm response and examine appropriate best practices for the management of alarms as safeguards.

1 Introduction

Almost all facilities will credit humans as safeguards, normally in the form of an operator responding to an alarm. Taking advantage of operators present at a facility is a very effective strategy - when implemented correctly - as they can perform numerous tasks, see patterns, learn causes and effects, and evaluate potential situations on the fly.

However, despite these advantages, operators are still human. Facilities must ensure that proper management goes into the use of humans as risk-mitigating safeguards, ensuring they are able to perform their actions in any scenario. Every safeguard in a facility should be carefully managed, from basic control loops to robust relief devices. The human safeguard should be no different.

Figure 1. Common safeguards breaks down the prevalence of different safeguards claimed at a real-life, complex petrochemical facility. As shown, alarms account for more than a quarter of all the claimed safeguards at the facility. Note that the percentage of alarms used as safeguards at this facility is lower than what is typically seen, as Safety Instrumented Systems (SIS) will not normally account for such a large percentage.

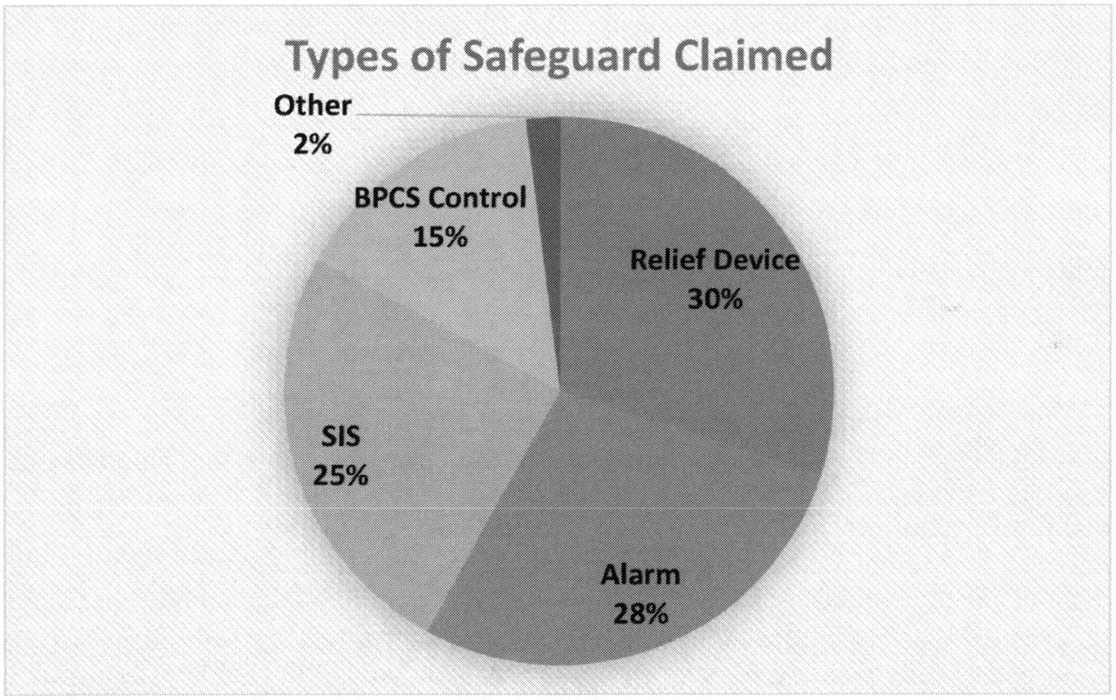

Figure 1. Common safeguards claimed at a petrochemical facility

When a Process Hazard Analysis (PHA) team takes credit for an alarm as a safeguard, the facility tends to overlook the entire safeguard which should include human response to the alarm. The alarm is thought of as the safeguard, not the human response. Since responding to alarm is an inherently human process, alarms and alarm management must always consider the human actions that must be taken. As such, facilities must ensure that an operator has adequate time to respond, knows what the correct response is, and will be able to troubleshoot a multitude of alarms to determine the correct one to answer.

Members of a PHA will often assume that alarms have correct setpoints and that operators will quickly know what to do in an alarm situation. Even with the requirement for operators participating in a PHA, little time is normally spent discussing the setpoint or if a human response is feasible. Additionally, many facilities do not consider checking the feasibility of responding to an alarm – it is often taken on good faith of an operator saying they think they can respond to it in time.

No safeguard claimed as frequently as an alarm is treated this way: relief devices must have extensive calculations showing they are adequately sized, and SIS have numerous requirements to prove their effectiveness. Provenance's experience with PST calculations, has shown that when reviewing adequate time to respond to an alarm, approximately 10-15% of alarms will have clear deficiencies. In many large facilities, thousands of alarms are claimed as safeguards, meaning that hundreds of alarms might be deficient, and an even greater number of hazardous scenarios might not be adequately mitigated.

Alarms – and the humans responding to them – can make extremely powerful safeguards, and they should not be removed as effective tools for mitigating hazardous scenarios. Rather, when claiming an alarm as a safeguard, facilities must pay more attention to ensure that the humans responding will be able to prevent an incident by ensuring that humans have adequate time to respond, know what the response should be, and know which alarm to respond to.

2 Overview of Process Hazard Analysis

2.1 Background on PHAs

While PHAs can be performed on any process, regardless of regulation, PHAs are required by all processes covered by OSHA's Process Safety Management Standard (PSM)[1] or EPA's Risk Management Plan (RMP) for Program Level 3 facilities.[2] As described by OSHA, a PHA is:

> [A]n organized and systematic effort to identify and analyze the significance of potential hazards associated with the processing or handling of highly hazardous chemicals.[3]

PHAs are the most common way for facilities to identify various safeguards for the safety, environmental, and economic hazards a facility might face. PHAs are conducted using a variety of methodologies from simple qualitative reviews (Checklists, What-ifs) to robust quantitative analyses (Fault Trees, Event trees). It should be noted most methodologies are devoted to identifying potential hazards and consequences, and there are additional methodologies commonly used in industry for quantifying risk, such as Layers of Protection Analysis (LOPA).[4]

Each of these methodologies and risk assessments have their own specific terminology but function similarly. This paper will focus on terminology from a common methodology (Hazards and Operability Study - HAZOP) and a common risk assessment method (LOPA), but almost all the ideas presented in this paper could be applied to other methodologies or assessments with minimal modification.

2.2 Key Assessment Review

2.2.1 Hazards and Operability Study (HAZOP)

The HAZOP Study is one of the most common PHA methodologies. In short, a HAZOP study follows a team evaluating potential causes of different deviations – a combination of a guide word, such as "No" or "Less", and a parameter like "flow" – and discussing the consequences from those deviations.[5] Safeguards for each cause-consequence pair are then discussed by the team, and it is determined if the safeguards are adequate or if action is needed to further mitigate risk.[6]

An example would be evaluating the closure of a vessel's vapor outlet valve in the "No/Less Flow" deviation and determining if the upstream vessel would be overpressured by the upstream pressure source. The team might examine a pressure alarm or relief valve present on the vessel as potential safeguards. Alarms, and their human responses, are very commonly credited safeguards. Because of the HAZOP's qualitative nature, it is often supplemented with a more quantitative risk assessment, especially for scenarios with higher risk consequences.

2.2.2 Layers of Protection Analysis (LOPA)

The Layers of Protection Analysis is recommended by the Center for Chemical Process Safety (CCPS) for a quantitative order of magnitude risk assessment methodology.[7] The LOPA risk assessment is widely used in industry to evaluate the risks of scenarios identified in PHAs, often combined with a HAZOP study to first identify the cause-consequence pairs that would be examined in a LOPA.

The safeguards identified as part of the HAZOP study are judged to see if they meet the criteria to be Independent Protection Layers (IPLs). IPLs are defined as having three characteristics: *independence* from the cause and other IPLs, *effectiveness* in preventing the undesired consequence from occurring or mitigating the unsafe aspects of the consequence, and *auditability* of testing or proving the IPLs effectiveness.[8] All IPLs are safeguards, but many safeguards will not meet the requirements to be an IPL.

Just as alarms and human actions are common safeguards, they are also common IPLs. The Probability of Failure on Demand (PFD) is used in industry to quantify how often the protection layer will fail. Human action IPLs normally are credited with a PFD of 1×10^{-1}/yr.[9] While this PFD might seem relatively low, this is equivalent to a SIS with a Safety Integrity Level (SIL) of 1, as defined by ANSI/ISA-61511, Functional Safety – Safety Instrumented Systems for the Process Industry Sector.

2.3 Where Humans are Credited in PHAs

Humans and their actions are present throughout most PHAs, as they are often the cause of a majority of considered scenarios and - as discussed in the assessment review - human responses to alarms are regularly considered as safeguards or IPLs. This paper will focus on the elements of human responses to alarms that would allow them to meet the requirements of an IPL, ensuring the human response can be independent, effective, and auditable.

While a facility might not use the LOPA process or declare alarms and their human response as IPLs, it is certainly in a facility's best interest to ensure any human safeguards are independent, effective, and auditable.

3 Human Safeguard Effectiveness

For operators to effectively respond to an alarm, several key factors must be met. A few depend on the alarm functioning properly – and will not be discussed in depth here – but there must be an acceptable amount of time to respond, appropriate procedures describing what to do in an alarm, and proper alarm prioritization for multiple alarms.

3.1 Operator Response Time

The amount of time for an operator to respond to alarm and prevent a hazardous event from occurring is similar to a term from ANSI/ISA-61511 for SIS, Process Safety Time. Process Safety Time (PST) is defined as:

> [the] time period between a failure occurring in the process or the basic process control system (with the potential to give rise to a hazardous event) and the occurrence of the hazardous event…[10]

ANSI/ISA-61511 further notes this is a property of the process itself,[11] meaning it does not consider the time for any actions to occur to prevent the hazardous event or lag time. Thus, any response time must be less than process safety time to be considered adequate. Note that while the ANSI/ISA-61511 definition of PST includes a time between failure and detection, this time will be neglected for the purposes of these calculations, as the time from failure to detection will not affect the operator's response.

A simplified timeline showing PST as a combination of lag time and the maximum allowable response time is shown in Figure 2.

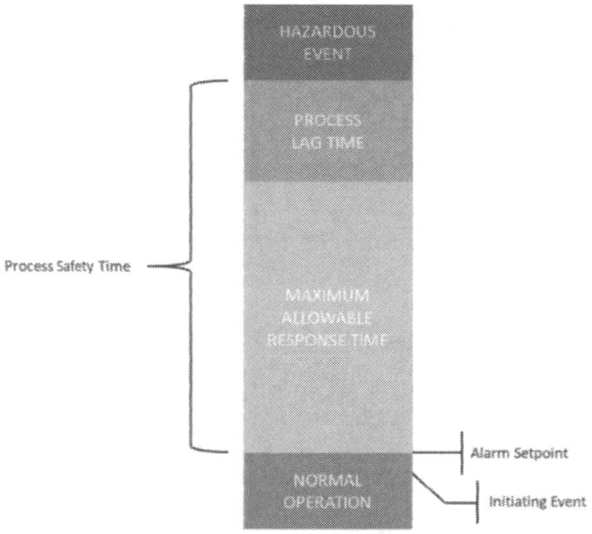

Figure 2. Process Safety Time and Maximum Allowable Response Time

Determining the allowable response time to an alarm – or any safeguard – provides an important criterion to judge the effectiveness of the operator response to an alarm.

Calculating the PST for a scenario can be performed in a variety of ways depending on the scenario in question. An overpressure, overfilling, or draining scenario might involve very simple calculations with mass imbalances and linear rates of change, while a PST for a runaway reaction would likely involve more complicated work such as reaction modeling and complex rates of change. Furthermore, PST can be estimated based on historical data of deviations in a process.[12]

As discussed previously, a true determination of the maximum response time an operator would have for a specific scenario would be less than the PST, as the lag time would need to be considered; however, the calculation of the PST can always be used as a first pass determination on the potential adequacy of an operator's response to an alarm. A PST of less than 5 minutes would likely disqualify the action regardless of any calculation of lag time, while a PST greater than 2 hours would very likely allow for plenty of time for operator response even with significant process lag.

3.2 Example System

An example system for a feed surge drum is presented in Figure 3. This system will be used as a reference point for discussions presented in this paper.

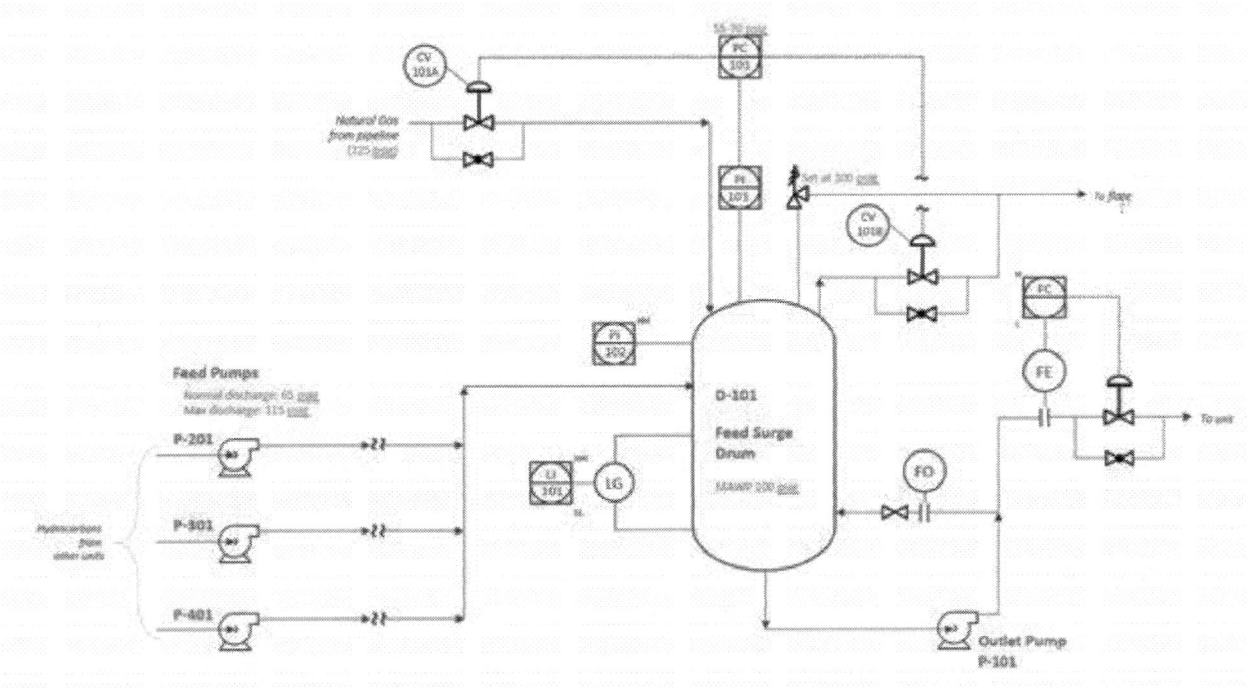

Figure 3. Feed Surge Drum

The Feed Surge Drum (D-101, MAWP of 100 psig) is fed with a hydrocarbon by three pumps from other units (P-201, P-301, and P-401). The drum has a blanketing system pulling natural gas from a pipeline that operates at 225 psig. A pressure controller on the drum controls the inlet blanket valve (CV-101A) and the outlet vent valve (CV-101B) to maintain an operating pressure in the drum between 55-70 psig. The liquid hydrocarbon leaves the system via P-101 and feeds the downstream unit. The surge drum has a high-level alarm, a high-pressure alarm (independent of the Pressure controller), and a relief valve set at 100 psig that relieves to flare.

Table 1. Example PHA scenarios, with alarms highlighted

Deviation	Cause	Consequence	Safety - Severity	Potential IPLs
1. No/Low Flow	**1.1** CV-101B malfunctions closed	**1.1.1** Blockage of vapor outlet. Continuing flow of natural gas, potential to exceed MAWP of the D-101 (100 psig) by upstream natural gas (225 psig). Potential for gasket failures/failure of the vessel, loss of containment, and potential fire.	High	PSV-101 set at 100 psig, relieving to flare
				PAHH-102 alarms at 80 psig, alerts operations to high pressure in the D-101. Operations opens vent valve to reestablish pressure in D-101
2. High Flow	**1.2** Misaligned bypass valve around CV-101A OPEN	**1.2.1** Increased flow of fuel gas into D-101. Bypass valve is significantly larger than CV-101A or CV-101B. Potential for accumulation of natural gas in D-101, potential for overpressure. Potential to exceed MAWP of the D-101 (100 psig) by upstream natural gas (225 psig). Potential for gasket failures/failure of the vessel, loss of containment, and potential fire.	High	PSV-101 set at 100 psig, relieving to flare
				PAHH-102 alarms at 80 psig, alerts operations to high pressure in the D-101. Operations opens vent valve to reestablish pressure in D-101

Table 1 above shows some scenarios the PHA team examines around the D-101 natural gas blanketing system. The PHA team first evaluates the natural gas blanket system for the drum being blocked on the outlet – consequence 1.1.1. The team determined the safety severity should be "High" when comparing the high upstream pressure of the natural gas (225 psig) to D-101 MAWP (100 psig). When examined in LOPA, the relief valve – which is adequately sized – will not reduce the potential frequency enough to meet the facility's internal guidance. Thus, the team selects the high-pressure alarm as the second IPL. The team agrees these IPLs are adequate; given the low normal flow rate of natural gas into the system, the operator response time is likely adequate.

The next scenario the team evaluates is the consequence of the bypass around CV-101A being opened. Given the upstream pressure is the same as the previous scenario and since the bypass is several sizes larger than CV-101A/B, it is determined the severity is the same. Once again, while the relief valve is adequately sized for this scenario, an additional IPL is required to fully mitigate the scenario per the site's guidance.

The team considers the high-pressure alarm on the drum as the next IPL. However, because the bypass around the control valves is much larger than the control valves and offers much less frictional resistance, the flowrate into the system is much higher and the alarm might not provide adequate time to respond.

To screen the potential IPL for consequence 1.2.1, the team performs a basic calculation of the PST using the net flowrate into D-101 (1922.8 lbm/hr of methane) with the bypass open, the vapor volume of D-101 (200 ft^3), the temperature of the natural gas (100°F), and the ideal gas law (a more robust equation of state could be used for higher accuracy). The starting point of the calculation is the alarm setpoint, 80 psig, while the final point is the MAWP of the vessel, 100 psig. Note that the final point should be the lowest point where no consequence would develop, however the point could be set even lower to prevent a SIS or relief device from actuating.

Using the above, it can be determined how the pressure changes as a linear function of time based on the net amount of natural gas as seen in the equations below:

$$(1)\ n_{consequence} = \dot{n}(PST) + n_{alarm}$$

$$(2)\ PST = \frac{(n_{Consequence} - n_{alarm})}{\dot{n}}$$

$$(3)\ n_{consequence} = \frac{P_{Consequence}V_{vapor\ Space}}{RT}, n_{alarm} = \frac{P_{alarm}V_{vapor\ Space}}{RT}$$

$$(4)\ PST = \frac{(P_{Consequence} - P_{alarm})V_{vapor\ Space}}{\dot{n}RT}$$

An engineer determines this simplified PST (20 seconds) is significantly less than the facility's guidance (30 minutes), and thus the operator response to the alarm would not be adequate.

Rather than relying on the alarm, the PHA team makes a recommendation for an administrative control to be placed on the oversized bypass valve to decrease the frequency of it being potentially mis-operated.

Utilizing a simplified PST screening can allow for quick determinations on the effectiveness of an operator response to an alarm for many scenarios. Depending on the type of facility, Provenance's experience with PST calculations has shown about 10-15% of previously credited IPLs have been found to have deficient response times. Additionally, most of the PSTs that would need to be evaluated fall into the more simplistic modeling cases (overfilling, overpressure, etc.) with linear rates of changes, meaning that adding this sort of screening can be relatively inexpensive and quick.

The team's review did note an operator would have adequate time to respond to the blocked vapor outlet case; however, an adequate time to respond does not mean that an effective response will occur. An operator must be able to easily diagnose the problem and perform the correct actions to mitigate the event.

3.3 Procedures for Abnormal Conditions

Ultimately, determining if the operator has time to respond to an alarm is pointless without the operator understanding the corrective actions that need to be taken if that alarm begins to sound. This can be accomplished by ensuring there are accurate procedures in place for alarm conditions and ensuring there is proper training involved on how to respond to these procedures. Procedures can be well written, but without training, an operator is more likely to make a mistake in the higher risk environment of alarm conditions.

Industry has made great strides in establishing solid procedures for normal operations; but - per OSHA - procedures are still one of the most common citations for PSM and have been since the inception of the program.[13] Many of these citations are based on a lack of procedures for abnormal conditions. Procedures should be established which describe how an operator is supposed to respond in abnormal conditions.

Returning to the previous Feed Surge Drum example, shown in Figure 3. Feed Surge Drum, it was noted during the PHA that the high-pressure alarm could be used for multiple scenarios including blocked liquid outlet, blocked vapor outlet, misalignment of an inlet valve, and failure of the pressure blanketing system.

The PHA team (or an alarm rationalization team) should discuss how the operator should respond to the high-pressure alarm, considering the multiple scenarios that could be encountered. The team agrees opening the manual vent valve to the flare would be adequate to relieve the excess pressure from all these scenarios. This is kept as part of the alarm procedures the operator should have access to and be trained on.

The alarm procedure for an individual alarm does not need to be a full-fledged document (unless the response is complicated enough to warrant one), but an operator should have procedures that state how to operate during abnormal conditions.

While it is often useful to reference a single alarm for multiple scenarios – to reduce the number of safeguards or IPLs to be maintained – it should be noted that when the operator receives an alarm, they will not be aware which scenarios are causing the issue without further investigation. This should always be kept in mind when deciding which safeguards should be used and how to properly develop a procedure for abnormal conditions. Additionally, any PST determination for an alarm should account for some amount of time for the operator to troubleshoot and research the problem to determine the correct action to take. This time will depend on the complexity of the process and the likelihood other alarms are sounding at the same time.

Having proper procedures in place is a key component of alarm management, but if operators are not trained in abnormal situations then alarm response will still be an issue. Operators will frequently cite a lack of comfort regarding the procedures around abnormal situation and will state additional experience and training is needed for these scenarios.[14] Since the tasks are related to abnormal conditions, normal "on the job" training often misses these scenarios.

The availability of simulations and other reliable means of hands-on training for abnormal situations can ensure operators have the confidence to adequately respond to alarm situations. Many facilities have training programs, but not enough training is dedicated to potential abnormal situations. Facilities should provide additional training especially focused on these situations.

Finally, when considering the human elements of any safeguard or IPL, it is important to remember that the complexity of the actions to be taken, the training (or lack of training) the operators possess, and the overall stress of the situation will greatly affect whether a human response will be adequate. New – or poorly trained – operators can increase the probability of error ten-fold or more in a high stress situation, and complicated procedures can lead to an increase of three-times the normal error rate.[15]

3.4 Alarm prioritization

An operator with adequate training, time to respond, and procedures still cannot fully be considered a perfect safeguard. While the previous sections have established the items needed to ensure that an individual alarm state and its response are acceptable, abnormal conditions are rarely accompanied by a single alarm.

Consider the system shown in Figure 3. Feed Surge Drum. If the outlet of P-101 is blocked, the Feed Surge will begin to overfill and overpressure. The control room operator would most likely begin to be assaulted by a wide range of alarms. The high-pressure alarm, the high-level alarm, and the low-flow alarm shown in the example would all sound. Since the feed to the unit is blocked, numerous downstream alarms would actuate as well; add in the large numbers of alarms upstream of D-101 which would also trigger to show an abnormal condition. While the response to the original alarm could be relatively simple – reestablishing flow from P-101 – the operator might not be able to even determine which alarm to prioritize if the entire board is going off.

This state of too many alarms, called alarm flood, is a common concern in industry where alarms have become extensive and part of all major processes.[16] As illustrated above, it is relatively easy to get large quantities of alarms sounding at the same time through minor changes to the process. One of the best ways to mitigate these alarm floods is to ensure a proper alarm rationalization is held. Alarm Rationalization, as described in ANSI/ISA-18.2, describes the process of ensuring alarms have proper prioritization, among other items.[17] By prioritizing alarms properly, an operator will be able to see which alarms need to be responded to first to properly mitigate the system. For example, part of this can be accomplished by highlighting the alarms that deal with personnel safety and assigning a higher priority to those that lead to the high severity consequences defined by the PHA.

In fact, since much of the data involved in a PHA and alarm rationalization are related to one another, some facilities have found use in conducting alarm rationalization processes and the PHA for a process at similar times. Many of the preliminary data gathering steps, determination of consequences without safeguards, and determining severity of the consequences are similar across the two processes. This can provide cost savings in terms of the preliminary steps and can help focus on the most important alarms to be managed. Most facilities that have considered this path have used it for new construction projects, but it could be applied to the existing paradigm to potentially increase cost savings.

Regardless of when the alarm rationalization process is conducted, a robust rationalization and alarm management program will help ensure alarms are properly prioritized to assist operators in making quick decisions needed to be an effective safeguard and help keep the process running smoothly.

4 Human Safeguard Independence and Auditability

While most of this paper deals with the effectiveness of the human element of safeguards, IPLs must also be independent of the cause, consequence, and other safeguards present. Additionally, an IPL must be auditable to validate the effectiveness of the safeguard via testing, documentation, review, or some other method.[18]

4.1 Independence

Compared to some of the more complex control loops which can be present as safeguards, it can be relatively easy to determine the independence of an alarm and the operator responding to it. In the alarm's case, it should be separate from any control loops that might be a part of the cause.

In our example, Figure 3. Feed Surge Drum, if CV-101A were to fail open or CV-101B were to fail closed, any alarm present on the common pressure controller would not be independent of the cause and thus would fail the independence determination. Note that CCPS's guidance for determining the independence of IPLs that route through a distributed control system (DCS) presents several approaches, as technically the failure of a logic solver could be a common cause failure between a cause and an IPL or two IPLs, even if they have different sensors and final elements.[19] Facilities can use one of the approaches described by CCPS, and one of which assumes IPLs are independent if they do not share a common sensor or final element.

The independence of the human element is also something that needs to be examined. To be independent of a common cause human failure (such as incapacitation), each alarm needs to have a separate human element as well. Often, facilities will only consider a maximum of one operator response to an alarm for a consequence, except in very specific scenarios – like consequences that would show up in multiple control rooms for instance. Additionally, depending on the action required, multiple operators might be required such as a board operator and a field operator. While there might be multiple field operator's prepared to manipulate a valve in the field, if there is only one board operator, common cause failures could prevent claiming multiple IPLs, as a failure of the board operator or issues in the control room might prevent multiple alarm responses from being carried out.

4.2 Auditability

The auditability of the human element of safeguards can sometimes appear to be a more complicated task than it is for instrumented safeguards which have a clear method of calculating the effectiveness of each and determining independence. Operator response times can be tested through drills; but with most facilities having thousands of alarms, this can be time intensive. However, as discussed earlier in this paper, there are simple steps that can be used to determine the effectiveness and independence of the alarm response and human elements of the safeguard.

Performing a calculation for PST, determining the response to the alarm state, and proper prioritization of the alarms should be documented in a similar way to other IPLs. Additionally, the logic behind the decisions and the assumptions must be documented as well so it is obvious how the decisions were made. Proper maintenance of these documents and the assumptions will allow for auditability of the operator response as a safeguard.

When performing training on different alarm states that can occur, the performance results from these trainings and any operator feedback can be incorporated with these documents to show the functions of the IPL are valid.

5 Additional Human Factors

While most of this paper has focused on the effectiveness of humans as a safeguard during an individual alarm state, there are many other human factors that can increase the effectiveness of operators and improve the operation of the plant. An in-depth review on operator error will not be presented but is important to discuss as these factors affect the performance of operators in all situations, including in abnormal conditions.

A facility should consider the design of the control room and Human-Machine Interface (HMI) in a way to reduce distraction, minimize fatigue, and ensure well-trained operators have the capability to perform in a proactive – rather than a reactive – manner. Ultimately, an alarm sounding should be treated as indication of some sort of impending failure, not relied on as a reminder to perform a task. As facility alarms and control systems have increased, the burden on operators has increased as well (the "paradox of automation").[20] To meet these increased demands, operators have naturally moved from a proactive state to a more reactive state. This change has occurred through no fault of the operators themselves, who can sometimes be perceived as lazy because of this shift.

Another aspect where the human element comes into play is through the PHA process itself. Humans (and operators) are parts of the PHA team and thus bring their own biases and experiences to the process. Social sciences have examined the relationship between a human's experience and their ability to forecast certain scenarios and their responses.[21] More often, humans will overestimate their own ability, especially in situations that they have not experienced. The lack of situational training for abnormal scenarios[22] means that operators will often be overconfident about their ability to respond to alarms accurately and rapidly. Additionally, facilities can foster a poor culture where safety becomes focused on "checking the box" and operators accept previously proposed safeguards assuming they have been vetted already.

A proper safety culture empowers operators by providing them with appropriate training, procedures, ergonomics, incident investigations, and support so they can work effectively in alarm states and potentially prevent alarms from even occurring. The savings from this sort of culture are obvious; by operating within the range of the process for more time, there is less downtime, decreased maintenance, and decreased hazardous situations.

6 Conclusion

Human safeguards (normally in response to an alarm) are some of the most widely used safeguards in the process industry; however, they often do not receive the same attention that SIS, interlocks, and other fully instrumented safeguards receive. Facilities that do not properly examine their alarms and the human responses to them could have many deficient safeguards. By examining the human responses as another part of the management of safeguards, facilities can better determine if they are adequately protected, if there are enough – or too many – alarms, and is there are places where they can improve the performance of the operators and the facility.

7 References

[1] 29 C.F.R. § 1910.119 (e)

[2] 40 C.F.R. § 68.67

[3] U.S. Occupational Safety and Health Administration, OSHA 3133: Process Safety Management Guidelines for Compliance, https://www.osha.gov/Publications/osha3133.html, Accessed on February 4, 2021.

[4] CCPS. Layer of Protection Analysis: Simplified Process Risk Assessment. Center for Chemical Process Safety, American Institute of Chemical Engineers, New York, NY, 2001.

[5] IEC 61882, Hazard and Operability studies (HAZOP studies) - Application guide, International Electrotechnical Commission, Geneva, Switzerland, 2016

[6] Ibid.

[7] CCPS. Layer of Protection Analysis: Simplified Process Risk Assessment. Center for Chemical Process Safety, American Institute of Chemical Engineers, New York, NY, 2001.

[8] Ibid, p. 80-88

[9] Ibid, p. 103

[10] Ibid, p. 23

[11] Ibid, p. 23

[12] G. Barnard and W. Creel, Impacts of Process Safety Time on Layer of Protection Analysis, Presented at the 11th Global Congress on Process Safety 2015 Spring Meeting, Austin, TX, 2015.

[13] I. Nimmo, Operator Effectiveness and the Human Side of Error, User Centered Design Services, Austin, TX, 2015.

[14] P.T. Bullemer and D.V.C. Reising, Managing Human Reliability: An Abnormal Situation Management Historical Perspective, Presented at the 18th Annual Internal International Symposium of the Mary Kay O'Connor Process Safety Center, College Station, TX, 2015, p. 7.

[15] CCPS. Guidelines for Initiating Events and Independent Protection Layers in Layer of Protection Analysis. Center for Chemical Process Safety, American Institute of Chemical Engineers, New York, NY, 2015, Appendix A.

[16] P.T. Bullemer and D.V.C. Reising, Managing Human Reliability: An Abnormal Situation Management Historical Perspective, Presented at the 18th Annual Internal International Symposium of the Mary Kay O'Connor Process Safety Center, College Station, TX, 2015, p. 8.

[17] ANSI/ISA-18.2, Management of Alarm Systems for the Process Industries, International Society of Automation, Pittsburgh, PA, 2016.

[18] CCPS. Layer of Protection Analysis: Simplified Process Risk Assessment. Center for Chemical Process Safety, American Institute of Chemical Engineers, New York, NY, 2001. p. 80

[19] Ibid, p. 86
[20] L. Bainbridge. Ironies of Automation, *Automatica*, vol. 19 (1983), p 775-779.

[21] D. Epstein. Range: Why Generalists Triumph in a Specialized World. Riverhead Books, New York, NY 2019. Focus on Chapter 10 "Fooled by Expertise

[22] P.T. Bullemer and D.V.C. Reising, Managing Human Reliability: An Abnormal Situation Management Historical Perspective, Presented at the 18th Annual Internal International Symposium of the Mary Kay O'Connor Process Safety Center, College Station, TX, 2015, p. 7.

AUTHOR INDEX

AIChE
120 Wall Street, FL 23
New York, NY 10005-4020

ISBN 978-1-7138-5330-5

CURRAN ASSOCIATES INC.
proceedings.com